Interpreting and Reporting Radiological Water-Quality Data

By David E. McCurdy, John R. Garbarino, and Ann H. Mullin

Book 5, Laboratory Analysis
Section B, Methods of the National Water Quality Laboratory
Chapter 6

Techniques and Methods 5–B6

U.S. Department of the Interior
U.S. Geological Survey

U.S. Department of the Interior
DIRK KEMPTHORNE, Secretary

U.S. Geological Survey
Mark D. Myers, Director

U.S. Geological Survey, Reston, Virginia: 2008

For product and ordering information:
World Wide Web: http://www.usgs.gov/pubprod
Telephone: 1-888-ASK-USGS

For more information on the USGS—the Federal source for science about the Earth,
its natural and living resources, natural hazards, and the environment:
World Wide Web: http://www.usgs.gov
Telephone: 1-888-ASK-USGS

Suggested citation:

McCurdy, D.E., Garbarino, J.R., and Mullin, A.H., 2008, Interpreting and reporting radiological water-quality data: U.S. Geological Survey Techniques and Methods, book 5, chap. B6, 33 p.

Contents

Figures

Tables

Conversion Factors

To Convert	To	Multiply by	To Convert	To	Multiply by
years (y)	seconds (s)	3.16×10^7	s	y	3.17×10^{-8}
	minutes (min)	5.26×10^5	min		1.90×10^{-6}
	hours (h)	8.77×10^3	h		1.14×10^{-4}
disintegrations per second (dps)	becquerels (Bq)	1.0	Bq	dps	1.0
Bq	picocuries (pCi)	27.0	pCi	Bq	3.7×10^{-2}
Bq/kg	pCi/g	2.70×10^{-2}	pCi/g	Bq/kg	37
Bq/m^3	pCi/L	2.70×10^{-2}	pCi/L	Bq/m^3	37
Bq/m^3	Bq/L	10^3	Bq/L	Bq/m^3	10^{-3}
microcuries per milliliter (μCi/mL)	pCi/L	10^9	pCi/L	μCi/mL	10^{-9}
disintegrations per minute (dpm)	μCi	4.50×10^{-7}	mCi	dpm	2.22×10^6
	pCi	4.50×10^{-1}	pCi		2.22
Tritium Unit (TU)	Bq/L	0.118	Bq/L	TU	8.47
	dpm/L	7.08	dpm/L		0.141
	pCi/L	3.19	pCi/L		0.313
cubic feet (ft^3)	cubic meters (m^3)	2.832×10^{-2}	cubic meters (m^3)	cubic feet (ft^3)	35.31
gallons (gal)	liters (L)	3.78	liters	gallons	0.265
gram (g)	ounce, avoirdupois (oz)	0.03527	ounce, avoirdupois (oz)	gram (g)	28.35
kilogram (kg)	pound (lb)	2.205	pound (lb)	kilogram (kg)	0.454

Acronyms and Abbreviations

α	probability of a Type I error
β	probability of a Type II error
λ	decay constant
CSU	combined standard uncertainty (1-sigma)
DQO	data quality objective
USEPA	U.S. Environmental Protection Agency
L	liter
L_C	critical level
MARLAP	Multi-Agency Radiological Laboratory Analytical Protocols Manual
MDC	minimum detectable concentration
MQO	measurement quality objective
NWIS	National Water Information System
NWQL	National Water Quality Laboratory
pCi	picocurie
PWS	performance work statement
QC	quality control
SI	International System of units
ssL_C	sample-specific critical level
ssMDC	sample-specific minimum detectable concentration
USGS	U.S. Geological Survey
WSC	Water Science Center

Interpreting and Reporting Radiological Water-Quality Data

By David E. McCurdy,[1] John R. Garbarino,[2] and Ann H. Mullin[2]

Abstract

This document provides information to U.S. Geological Survey (USGS) Water Science Centers on interpreting and reporting radiological results for samples of environmental matrices, most notably water. The information provided is intended to be broadly useful throughout the United States, but it is recommended that scientists who work at sites containing radioactive hazardous wastes need to consult additional sources for more detailed information. The document is largely based on recognized national standards and guidance documents for radioanalytical sample processing, most notably the Multi-Agency Radiological Laboratory Analytical Protocols Manual (MARLAP), and on documents published by the U.S. Environmental Protection Agency and the American National Standards Institute. It does not include discussion of standard USGS practices including field quality-control sample analysis, interpretive report policies, and related issues, all of which shall always be included in any effort by the Water Science Centers. The use of "shall" in this report signifies a policy requirement of the USGS Office of Water Quality.

Introduction

Interpreting and reporting radiological results requires a full understanding of the concepts of detectability and quantification unique to radiochemistry and radiation emissions. Radioactivity describes a group of processes by which matter and energy are released from the nucleus of atoms as an alpha particle (^{4}He$^+$ nucleus), a beta particle (equivalent to an electron), or a gamma ray (photon or energy wave). The residual nucleus usually is transformed to a different element, and for alpha- and gamma-emitting nuclides, the energy of the emission(s) can be measured to identify the source radionuclide. Natural and anthropogenic radionuclides occur widely in the hydrologic environment (Hem, 1985, p. 146–151; Drever, 1988, p. 379–381; Lieser, 2001). The rate of decay of a given quantity of radioactive atoms ($-dN/dt$) is proportional to the amount of atoms present (N) as shown in equation 1 where λ is the decay constant.

$$-\frac{dN}{dt} = \lambda N \qquad (1)$$

The USGS currently (2008) reports the activity of a radionuclide in curies (Ci). The corresponding International System (SI) unit is the becquerel (Bq), and one curie equals 3.7×10^{10} Bq. The rate of decay for environmental samples is commonly measured in picocuries (pCi = 10^{-12} Ci). One curie is defined as 3.7×10^{10} disintegrations per second, the approximate rate of alpha radiation from one gram of radium. One pCi is thus 0.037 disintegration per second or 2.22 disintegrations per minute (dpm). Radiological analysis in essence involves detecting and counting individual decay-product emissions from a sample for enough time to compute an average rate. The decay constant λ is inversely proportional to the half-life of the radionuclide ($T_{1/2}$), which is the time required for half of the original amount to decay as shown in equation 2.

$$\lambda = \ln 2 / T_{1/2} \qquad (2)$$

A radionuclide concentration reported in picocuries per unit of measure can be converted to specific activity (SA) per moles or grams by using the half-life of the radionuclide and Avogradro's number using equation 3.

$$SA = \left(\lambda \times N_o \right) / 3.7 \times 10^{-2} = (N_o \times 18.73) / T_{1/2} \qquad (3)$$

where N_o is Avogradro's number of atoms (6.023×10^{23}) per gram mole of the radionuclide, λ is the decay constant (unit of 1 per second), and $T_{1/2}$ is half-life of the radionuclide in units of seconds. By incorporating Avogradro's number into equation 3, a simpler equation 4 can be derived:

$$SA = (1.13 \times 10^{25}) / (AW \times T_{1/2}) \qquad (4)$$

where AW is the atomic weight of the radionuclide. The following table provides similar equations for a specific activity (but in units of picocuries per gram, pCi/g) for radionuclides that have half-lives in years, days, hours,

[1]McCurdy Associates, Northboro, Mass.

[2]U.S. Geological Survey National Water Quality Laboratory, Denver, Colo.

$T_{1/2}$	Equation (pCi/g)	Equation (g/pCi)
Years	$SA = (3.58 \times 10^{17})/(AW \times T_{1/2})$	$AW \times T_{1/2} \times 2.80 \times 10^{-18}$
Days	$SA = (1.31 \times 10^{20})/(AW \times T_{1/2})$	$AW \times T_{1/2} \times 7.63 \times 10^{-21}$
Hours	$SA = (3.14 \times 10^{21})/(AW \times T_{1/2})$	$AW \times T_{1/2} \times 3.18 \times 10^{-22}$
Minutes	$SA = (1.88 \times 10^{23})/(AW \times T_{1/2})$	$AW \times T_{1/2} \times 5.32 \times 10^{-24}$
Seconds	$SA = (1.13 \times 10^{25})/(AW \times T_{1/2})$	$AW \times T_{1/2} \times 8.85 \times 10^{-26}$

minutes, and seconds. For example, by using the equation above corresponding to $T_{1/2}$ in years, 1 pCi of ^{238}U with a half-life of 4.46×10^9 years corresponds to a mass of 2.97×10^{-6} grams ($238 \times 4.46 \times 10^9$ years $\times 2.80 \times 10^{-18}$). A more complete introduction to environmental radiochemistry can be found in Eisenbud and Gesell (1997).

Some radiochemical analyses involve separation or isolation steps followed by a quantitation step typically based on some radiation emission measurement. The uncertainty of a radiological result is affected by the activity in the sample, the duration of the measurement, and other various factors that can be controlled for that single measurement. Supplemental information and guidance relative to these concepts have been provided in an Appendix.

Analytical laboratories that provide chemical, radiochemical, and biological analyses to the U.S. Geological Survey (USGS) shall be evaluated relative to the objectives of a project requiring analyses and approved for use for that specific project. Analysis of performance-testing samples will provide the basis for the initial laboratory approval, and an approved laboratory must continue to provide acceptable performance-testing sample results during the life of the project. The National Water Quality Laboratory (NWQL) submits performance-testing samples before the award of all radioanalytical contracts it administers. Review the Policy for the Evaluation and Approval of Analytical Laboratories before submitting samples for analysis (U.S. Geological Survey Office of Water Quality Technical Memorandum No. 2007.01, 2007). Additional information on the laboratory evaluation process can be found at *http://qadata.cr.usgs.gov/lep* (accessed 2008).

USGS National Programs and Water Science Centers (WSCs) conduct research studies and monitoring programs that focus on the detection and quantification of radiological constituents in various environmental matrices, most notably water. In order to support these studies and programs, the USGS NWQL maintains memoranda of understanding and contracts with various USGS and commercial laboratories to process and analyze samples collected by the WSCs. Within the memoranda of understanding and contracts are generic performance work statements (PWS) that apply to all USGS projects. Once the analyses have been completed, the contract laboratories provide NWQL with the results for each radiological constituent and sample matrix combination. The NWQL staff evaluates the reported data and information for technical issues and compliance to specifications stated in the PWS. The acceptable data are transferred to the USGS

National Water Information System (NWIS) database through which they are provided to WSCs for further analysis, interpretation, and publication. The ancillary data stored in NWIS include all the parameters needed to publish the data in a report or on NWISWeb including the result, its associated combined standard uncertainty, the sample-specific critical level, and appropriate remark and value-qualifier codes. An additional data package sent by the NWQL to the WSCs provides radiological results with their associated combined standard uncertainty and other parameters and data, which provide ancillary measurement information including quality-control sample results. These ancillary data are used by NWQL to make decisions about detection, uncertainty of results, and reporting of data. Definitions and use of these ancillary data and information are detailed elsewhere in this document.

The laboratories used by NWQL analyze environmental samples according to the PWS, including the measurement quality objective of the *a priori* minimum detectable concentration (*a priori* MDC) for a radiological constituent and sample matrix combination. Unacceptable Type II errors (false nondetection) are avoided through the establishment of the *a priori* MDC and the method selected by the laboratory. Limiting unacceptable Type I errors (false detection) is assured by verifying that each reported result exceeds its respective sample-specific critical level (ssL_C). Specifications for the Data Quality Objective (DQO) required detection levels for certain radiological constituents relative to the USEPA Safe Drinking Water Act (SDWA) also are included in the PWS. In this report, DQO is used in the same manner as the term Measurement Quality Objective used in the Multi-Agency Radiological Laboratory Analytical Protocols Manual (MARLAP). The *a priori* MDC is the only DQO specified by NWQL in the PWS. The PWS also will specify the approved radiological method that must be used for the analysis (U.S. Environmental Protection Agency, 1980a). Other important information specified in the PWS include the calculation of the ssL_C, sample-specific Minimum Detectable Concentration (ssMDC), and Combined Standard Uncertainty (CSU) of the measured result. The reported laboratory values for these measurement parameters are used by NWQL personnel to verify contractual compliance with PWS specifications, to determine detection of radiological constituents, to interpret the quality of the result, and to support decisions related to data usability and reporting of data in reports.

Appendix sections A3 and A4 contain the technical basis for the ssL_C and the ssMDC as defined in chapter 20 of the Multi-Agency Radiological Laboratory Analytical Protocols Manual (MARLAP, 2004) and a discussion on the effect of sample size and instrument sample processing time on achieving a specified minimum detectable concentration. Terms are defined in a Glossary at the back of the report.

1. Analytical Information Reported by the Contract Laboratory

Various information related to the sampling and radio-analytical processes are reported by the contract laboratories to NWQL and subsequently to USGS WSCs. An Electronic Data Deliverable (EDD) is provided to NWQL by the contract laboratory that contains the unrounded result, CSU, ssL$_C$, and ssMDC. The information to be reported by the contract laboratory is defined within the PWS and includes sampling and laboratory processing parameters. The analytical information includes:

- client sample identification (ID) code

- sample collection date

- sample matrix

- sample size received

- special instructions

The contract laboratory also reports the following information with the analytical results:

- laboratory identification number cross-referenced to client sample ID

- sample receipt date

- analyte (radiological constituent)

- analysis date

- result value (positive, negative, or zero)

- combined standard uncertainty (CSU; 1-sigma uncertainty)

- sample-specific minimum detectable concentration (ssMDC; *a posteriori* MDC)

- contract required minimum detectable concentration (MDC; *a priori* MDC)

- sample-specific critical level (ssL$_C$)

- chemical yield (percent) for radiochemical processing

- aliquant size processed

- tracer used

The contract laboratory provides additional information in the form of narrative comments for use in evaluating results for data usability. Also, the method used and its reference designation are provided. All of this information is included on the compact disk data package that is sent to the contact person listed on the Analytical Services Request form (see section 3.6).

2. Definitions of Important Analytical Parameters

2.1 Combined Standard Uncertainty

The Combined Standard Uncertainty (CSU) can be viewed as the statistical standard deviation of an individual radiological result. The concentration of a radiological constituent in a sample is typically calculated using a mathematical equation that includes such parameters as the measured signal response of a radiation detector (events per time unit), the detector background signal response, the detector efficiency for the radiation emission producing the response, sample aliquant size processed, chemical yield of the radiochemical process, and decay and ingrowth factors based on the half-life of the radionuclide or its decay product. Each measurement parameter in the equation has its own uncertainty defined as a standard uncertainty. The CSU of the final result is determined using the common statistical approach that the variance (squared CSU) of a function of several variables can be approximated by applying the function to the variance of each variable component (for example, Benjamin and Cornell, 1970, p. 180–186; MARLAP, 2004, chapter 19). Using this logic, the CSU of a radiological result is the square root of a sum of variances. The Appendix provides an example of a generic equation for calculating concentration (section A1) and the propagation of the standard uncertainties to derive the combined standard uncertainty (section A2).

The statistical normal distribution describes the uncertainty of most contributing variables in a radiological analysis, but the measured number of counts for an analysis follows the Poisson distribution. Poisson variables have a lower bound (zero for radiation counts) and thus have a positive skew. For sufficiently large counts, however, the Poisson distribution can reasonably be approximated as a normal distribution. All the statistical calculations in this report treat the uncertainty of a radiological result as a normal distribution.

When a concentration and its associated CSU are reported, a confidence interval can be calculated that defines the range of concentration (the lower and upper concentration) for the "true concentration" with a certain confidence. Contract laboratories calculate and report the CSU at the 68-percent or 1-sigma (1σ) confidence level (analogous to the standard confidence level used when reporting the standard deviation for other water-quality results). The confidence level that is used when interpreting or publishing radiological results is dependent on the DQOs of the project. Reporting the concentration with its corresponding CSU (as provided in the NWIS database) provides the 68-percent confidence interval. The WSC shall always state the level of confidence of the CSU that is reported; for example, 1.25 ± 0.25 picocuries per liter (pCi/L) at 1σ or 1.25 ± 0.25 pCi/L at the 68-percent confidence level. The corresponding 68-percent confidence interval would be 1.00 to 1.50 pCi/L; or in other words, there is a 68-percent chance that the true value is between 1.00 and 1.50 pCi/L.

For most radionuclide concentrations reported by NWQL, the principal contributor to the CSU is the standard uncertainty of the net count rate. The relation between the CSU and the calculated activity for two radioanalytical measurement techniques is shown in figures A5a, A5b, A6a, and A6b in section A4.5 of the Appendix.

2.2   Sample-Specific Critical Level (ssL$_C$)

The critical level (L$_C$) is the smallest measured concentration that is statistically different from the instrument background or analytical blank, and it serves as the detection threshold for deciding whether the radionuclide is present in a sample. The L$_C$ is calculated from measurements obtained using nominal or typical analytical parameter values, whereas the sample-specific critical level (ssL$_C$) is calculated from measurements obtained using the same analytical parameter values that were used during the analysis of a sample. USGS PWSs require the routine calculation of the ssL$_C$ for each sample, using parameter values that were actually measured during the generation of the sample result. The null hypothesis for establishing the critical level is that "the sample activity level is the same as the measured instrument background or blank sample value." The maximum acceptable probability α of false detection (significance level), together with the standard deviation of the net blank sample distribution having a mean value of zero, forms the basis for the critical level upon which detection decisions may be made (Currie, 1968). For analysis of USGS radiological samples, a false detection rate of 5 percent (α_0=0.05) is used. This hypothesis test strives to limit false detection (known as Type I error). Figure A1, section A3.1 in the Appendix graphically illustrates the critical level concept. Note that the critical level concept as applied to radionuclide detection is based on a "one-sided" hypothesis test that considers only the upper-tail probabilities of the null distribution and is different from a two-sided test that would consider both the upper- and lower-tail probabilities.

A detection decision is based on comparison of the sample result with the ssL$_C$. Because the ssL$_C$ is a hypothesis-testing concept based on a preestablished probability of false detection and the standard deviation of the net background distribution, the combination of the result and the ssL$_C$ and not the measurement CSU (and resultant symmetrical confidence interval) is used for detection decisions. Whenever the concentration of a radiological constituent is greater than the ssL$_C$, it shall be considered detected; that is, the reported concentration is positive and greater than the measurement (instrument) background or the radiological constituent's concentration in a blank sample. When the concentration is greater than the ssL$_C$, the decision "detected" shall be reported, and a symmetrical confidence interval shall be given "after the detection decision is made" (Currie, 1968).

Scientists can evaluate the reported sample data set to determine if the reported ssL$_C$ has been calculated properly. Section A3.3 of the Appendix provides a practical approach for verifying the ssL$_C$ for most radioanalytical methods using the reported CSU. This guidance is not definitive but may be used to determine whether or not the relation of reported ssL$_C$ and its uncertainty is reasonable.

2.3   Minimum Detectable Concentration

The critical level concept discussed in section 2.2 addresses Type I error (false detection), but it does not consider Type II error (false nondetection). If the true concentration of a radionuclide were exactly equal to the critical level, the inherent uncertainty of the measurement would produce larger (detected) results for some samples and smaller (not detected) results for others. The Minimum Detectable Concentration (MDC) concept addresses Type II error. The MDC can be calculated *a priori*, using nominal or typical analytical parameter values, or *a posteriori* for a specific sample, using the ssL$_C$ and parameter values for an individual sample.

2.3.1   *a priori* Minimum Detectable Concentration (*a priori* MDC)

Consideration of both Type I and II errors is the basis of the *a priori* MDC concept. The critical level incorporated in the expression for the *a priori* MDC (see equation A9 of section A4.1 of the Appendix) is typically calculated using nominal or typical parameter values such as detector efficiency, chemical yield, and sample aliquant processed. The *a priori* MDC for a radioanalytical method is a laboratory analytical-method performance characteristic.

The *a priori* MDC is an *a priori* (before the sample measurement) concept that is only used to facilitate comparisons of the relative detection capabilities of measurement systems or radiological methods. It is defined as the lowest true concentration that gives a specified probability that the measured concentration will exceed its critical level concentration (Currie, 1968; MARLAP, 2004, chapter 20). The *a priori* MDC satisfies the hypothesis testing probability (β) at the 0.05 or 95-percent confidence level that the true result is greater than the L$_C$, given that the analytical result equals the *a priori* MDC and assuming a normal distribution. As such, its definition may be restated as the lowest concentration for which there is a 95-percent probability of producing a result greater than the critical level and a 5-percent probability of falsely concluding that a blank measurement represents a positive measurement (above the critical level). Figure A2, section A4.1 in the Appendix, graphically illustrates the *a priori* MDC concept and shows a distribution of measurement results from a set of samples having a radiological constituent concentration at the MDC.

Based on USGS program needs and state-of-the art radioanalytical methods, standardized *a priori* MDC DQOs for various radiological constituent/matrix combinations have been established by NWQL. The *a priori* MDC for a radiological analyte is in the NWQL Catalog (see the reporting level entry at *http://nwql.cr.usgs.gov/usgs/catalog/index.cfm*). The established *a priori* MDC DQO specification is based on need or the expected detection capability of a method, or both. For example, an *a priori* MDC contractual specification of 1 pCi/L for tritium (³H) in water has been chosen for certain research studies. These standardized *a priori* MDC concentrations and matrix combinations become laboratory contract specifications as defined in the PWS. By establishing the *a priori* MDC, an acceptable Type II error is defined. When the laboratory selects a method to meet the *a priori* MDC, unacceptable Type II errors are limited.

The contract laboratory uses the *a priori* MDC requirements to select appropriate methods and method parameters to meet the contract specifications. Nominal or typical parameter values (detector efficiency, chemical yield, sample aliquant processed, and so forth) of the radioanalytical method are generally chosen by the contract laboratory when calculating the *a priori* MDC for a given method and radiological constituent.

2.3.2   Sample-Specific Minimum Detectable Concentration (ssMDC)

NWQL contracts also require calculation and reporting of an *a posteriori* (after the measurement) or sample-specific Minimum Detectable Concentration (ssMDC) in association with each radiological result reported. The contract laboratory uses the ssL_C as the basis for establishing the ssMDC as evaluated in context with the *a priori* MDC on each individual measurement to establish the Type II error. The ssMDC is used by the NWQL to verify that the *a priori* MDC DQO has been met. Because the ssMDC is calculated with actual parameter values used during the analysis of the sample in question, the ssMDC, in most cases, may tend to be below the *a priori* MDC, which uses more conservative nominal method parameter values. In most cases, the actual method parameter values used for the ssMDC calculation do not substantially change from the nominal values used in the *a priori* MDC calculation unless certain circumstances have occurred; for example, smaller sample size processed or lower chemical yields may lead to longer counting times. Occasionally, if the laboratory does not adjust certain method parameters, the *a priori* MDC DQO may not be met because of possible unexpected chemical and instrumental interferences and small sample sizes. In such cases, the ssMDC will be greater than the *a priori* MDC DQO.

The MDC (*a priori* MDC or *a posteriori* ssMDC) shall never be applied to make decisions about whether a radiological constituent has been detected in a sample; rather, the ssL_C shall be used for defining when a concentration is different from zero with a specified probability (5 percent for most cases) of false detection. Section A4.2 of the Appendix

provides additional information and the typical equations used by a laboratory to calculate the ssMDC. Practical approaches for determining whether a reported ssMDC has been calculated properly (section A4.3), and the effects of sample volume and counting time on the magnitude of the ssMDC (section A4.4) are provided in the Appendix.

2.4   Comparison of Radiological, Inorganic, and Organic Detection Levels

The concepts on radiological detectability as presented in section 2.2 on the ssL_C and section 2.3.2 on the ssMDC (also sections A3 and A4 in the Appendix) are similar to those presented for organic and inorganic analytes in the U.S. Geological Survey Open-File Report 99–193 (Childress and others, 1999) for the long-term method detection level (LT–MDL) and the laboratory reporting level (LRL), respectively. The LT–MDL is based on a modification of the U.S. Environmental Protection Agency's method detection limit (MDL) procedure and relies on several of its key assumptions (Childress and others, 1999). The primary difference between the MDL and the LT–MDL is that the LT–MDL is designed to measure more sources of variability and therefore is expected to be higher than the MDL. The MDL uses the standard deviation of spiked samples based on a minimum number of seven spiked samples (a snapshot or single set of measurements made at between 1 and 5 times the estimated MDL concentration), whereas the LT–MDL uses the standard deviation based on a much larger set of spiked samples (at least 24 per year) collected over an extended period of time, typically 6 to 12 months.

The basic concepts presented for radiological, organic, and inorganic analytes assume a normal distribution for the blank and spike sample measurements and use the standard deviations of these distributions and defined error rates for false detection and false nondetection. The equations for the critical level (L_C and ssL_C) and the long-term method detection level (LT–MDL) are basically identical except that the critical level equations use a 5-percent α error rate for false detection compared with 1-percent α error rate for false detection used to calculate the long-term method detection level. For both applications, the critical level and LT–MDL are incorporated into the determination of the MDC and the LRL, respectively. Similar to the L_C and the LT–MDL, the error rates for the MDC and LRL differ; the MDC equation uses a 5-percent β error rate for false nondetection, whereas a 1-percent β error rate for false nondetection is used for the LRL.

Another basic difference in calculating the L_C and the LT–MDL is that the L_C uses a standard deviation for a distribution of blank sample results whereas, for organic analytes and some inorganic analytes, the LT–MDL uses a standard deviation of a distribution of results from samples spiked near the estimated detection level. In both cases, the calculations assume the standard deviation in blanks and the standard deviation near the critical or detection level are equal. A comparison of the basic differences between detection-level terms and error rates and basic assumptions, detection decisions, and results reporting for radiological, organic, and inorganic constituents is presented in tables 1 and 2, respectively.

3.   National Water Quality Laboratory Evaluations of Contract Laboratory Results

Initial evaluation of the quality of contract laboratory radiological data is conducted by NWQL before the data are sent to the NWIS database. This initial evaluation is done largely to determine contractual compliance and overall quality of the data. The data are then transferred to the NWIS database through which they can be accessed by USGS WSCs.

3.1   Initial Evaluation Criteria

As a standard practice, NWQL evaluates a contract laboratory's reported radiological data for each sample for contractual compliance for technical items such as, but not limited to:

- reporting of sample parameters and information according to specifications (see section 1);

- contractual MDC specification (by comparing the ssMDC value to the contractual *a priori* MDC);

- radiological hold time (by comparing sample collection and analysis dates);

- processing turnaround time (by comparing sample receipt and analysis report dates);

- insufficient sample size for analysis;

- yield for certain radiological constituents; and

- batch Quality Control (QC) sample results related to method bias (laboratory control and matrix spike samples), excess uncertainty or imprecision (split or duplicate sample analyses), and false positive and negative (blank samples).

The typical processes that NWQL uses to evaluate contract laboratory results are listed in table 3. Several examples of the most common and relatively unambiguous situations and those that occur less frequently and require deeper scrutiny are discussed.

3.2   Rounding Results

The NWQL Laboratory Information Management System (LIMS) rounds contract laboratory results received through the EDD by using the American National Standards Institute procedure N42.23 (American National Standards Institute, 2003). The CSU shall be rounded to two significant figures, and both the radiological concentration and CSU shall be reported to the same number of decimal places. Proper rounding conventions notwithstanding, one must always remember that the CSU reported in association with the sample concentration, and not the base-10 rounding of results, establishes the number of significant digits in a radiological result. Examples are provided in table 3 showing how contract laboratory radionuclide concentrations and their CSUs are rounded before they are sent to the NWIS database.

Table 1.   Equations and error rates used for calculating the critical level and long-term method detection level.

[L_C, critical level; LT–MDL, long-term method detection level; MDC, minimum detectable concentration; LRL, laboratory reporting level; NA, not applicable; s_{blanks}, standard deviation for blanks; s_{LT-MDL}, standard deviation for spikes at 1 to 5 times the estimated detection level; k_β, statistical factor; $t_{(n-1,1-\alpha=0.99)}$, statistical factor; n, number of samples; α, probability of false detection; β, probability of false nondetection]

	$L_C{}^a$	LT–MDL[b]	*a priori* MDC[c]	LRL
Basic practical equation	$s_{blanks} \times k_\alpha$	$s_{LT-MDL} \times t_{(n-1,1-\alpha=0.99)}$	$L_C + k_\beta \times s_{blanks}$	$2 \times s_{LT-MDL} \times t_{(n-1,1-\alpha=0.99)}$
Specific equation	$s_{blanks} \times 1.645$	$s_{LT-MDL} \times 2.50$	$2.71 + 3.29 \times s_{blanks}$	$2 \times$ LT–MDL
False detection error rate	0.05	0.01	0.05	0.01
False nondetection error rate	NA	NA	0.05	0.01

[a]When $\alpha = 0.05$, $k_\alpha = z_{(1-\alpha)} = 1.645$; where $z_{(1-\alpha)}$ denotes the $(1-\alpha)$ quantile of the standard normal distribution.

[b]The basic equation is described by Childress and others (1999). The Student's t statistic for 23 degrees of freedom is equal to 2.50 for $\alpha = 0.01$.

[c]For practical purposes, the basic assumption is that s_{blanks} is approximately equal to $s_{spikes-MDC}$ (the standard deviation of the distribution of a sample spiked at the MDC). The equation simplifies by assuming $\beta = 0.05$ and $k_\alpha\beta = z_{(1-\beta)} = 1.645$; where $z_{(1-\beta)}$ denotes the $(1-\beta)$ quantile of the standard normal distribution.

Table 2. Basic assumptions, detection decisions, and results reporting for radiological, organic, and inorganic methods.

[L_C, critical level; ssL_C, sample-specific critical level; *a priori* MDC, method-specific *a priori* minimum detectable concentration; ssMDC, sample-specific MDC; LT–MDL, long-term method detection level; LRL, laboratory reporting level; PWS, performance work statement for contract laboratory; NA, not applicable; lc std, lowest calibration standard; ≥, greater than or equal to; <, less than]

Radiological methods	L_C	ssL_C
Basic assumptions	Typical blank or instrument background distribution and sample parameters; can be instrument specific or average value for all instruments using method	Instrument-specific background distribution and sample-specific parameters
Detection decisions	Not used	Result ≥ ssL_C
Reporting results	Not reported	Is always reported with the result (negative, zero, or positive) and the combined standard uncertainty (CSU)
Organic and inorganic methods	**LT–MDL**	**No corresponding term**
Basic assumptions	Multiple instrument and multiple analysts; analysis of samples spiked at 1 to 5 times the estimated detection level; spike distribution of ≥ 24 samples over 6 to 12 months; and constant sample parameters used	NA
Detection decisions	Result ≥ LT–MDL	NA
Reporting results	Result concentrations ≥ LT–MDL are reported with a qualifier when the concentration is less than the LRL or lc std, whichever is greater (ideally, the lc std is equal to the LRL); when the result < LT–MDL, then < LRL is reported; for information-rich organic methods, qualitative results are reported with a qualifier when a result is < LT–MDL[b]	NA
Radiological methods	***a priori* MDC**	**ssMDC**
Basic assumptions	Typical blank or instrument background distribution and sample parameters; can be instrument specific or average value for all instruments using method	Instrument-specific background distribution and sample-specific parameters
Detection decisions	Not used	Not used
Reporting results	Available from NWQL Catalog[a]	Not generally reported; it is used only to evaluate contractual requirements of the PWS
Organic and inorganic methods	**LRL**	**No corresponding term**
Basic assumptions	Multiple instrument and multiple analysts; analysis of samples spiked at 1 to 5 times the estimated detection level; spike distribution of ≥ 24 samples over 6 to 12 months; and constant sample parameters used	NA
Detection decisions	Not used	NA
Reporting results	All results greater than the LRL or lc std, whichever is greater, are reported without a qualifier[b]	NA

[a] The *a priori* MDC is listed in the NWQL Catalog under the reporting level entry; see http://nwql.cr.usgs.gov/usgs/catalog/index.cfm

[b] Refer to figure 10 in U.S Geological Survey Open-File Report 99–193 (Childress and others, 1999) for details

Table 3. Examples showing the processes that are used by the National Water Quality Laboratory to review radiological results.

[All numbers are in units of picocuries per liter; CSU, 1-sigma Combined Standard Uncertainty; ssL_C, sample-specific critical level sent to the National Water Information System (NWIS) database; a priori MDC, contractual a priori Minimum Detectable Concentration; ssMDC, sample-specific Minimum Detectable Concentration; Code, remark or value-qualifier code(s) sent to the NWIS database; R, nondetect, result less than sample-specific critical level;), ssMDC exceeded the a priori MDC; =, negative result may indicate potential negative bias; NWQL, National Water Quality Laboratory; ±, plus or minus; <, less than; >, greater than; WSC, Water Science Center]

Unrounded concentration	Unrounded CSU	ssL_C	ssMDC	a priori MDC	Code	Decision processes	Rounded concentration and (CSU) sent to the NWIS database
2.346	0.542	0.93	2.3	3.0		Result > ssL_C; reasonable relation between CSU, ssL_C and ssMDC; ssMDC < a priori MDC.	2.35 (0.54)
0.534	0.542	0.93	2.3	3.0	R	Result < ssL_C; reasonable relation between CSU, ssL_C and ssMDC; ssMDC < a priori MDC. Remark code sent to NWIS.	0.53 (0.54)
6.636	1.542	3.2	6.5	3.0	)	Result > ssL_C; reasonable relation between CSU, ssL_C and ssMDC. However, the ssMDC > contractual a priori MDC. NWQL will try to determine the cause by looking at other data and information and consulting with the contract laboratory. Reanalysis may be requested. Value-qualifier code sent to NWIS.	6.6 (1.5)
0.534	0.742	1.6	3.2	3.0	R)	Result < ssL_C; reasonable relation between CSU, ssL_C and ssMDC. However, the ssMDC > contractual a priori MDC. NWQL will try to determine the cause by looking at other data and information and consulting with the contract laboratory. Reanalysis may be requested. Remark and value-qualifier codes sent to NWIS.	0.53 (0.74)
-1.525	0.972	0.93	2.7	3.0	R	Result < ssL_C; reasonable relation between CSU, ssL_C and ssMDC; ssMDC < a priori MDC; and no negative bias (negative concentration < 1.65 × CSU). Remark code sent to NWIS.	-1.52 (0.97)
-0.504	0.735	0.22	0.70	0.90		The ssMDC < a priori MDC and there is no negative bias (negative concentration < 1.65 CSU). However, the result is unacceptable because the ssL_C and ssMDC are too small relative to the CSU. NWQL may request a recalculation or reanalysis.	Results are not sent to the NWIS database unless the problem is resolved.

Table 3. Examples showing the processes that are used by National Water Quality Laboratory to review radiological results.—Continued

[All numbers are in units of picocuries per liter; CSU, 1-sigma Combined Standard Uncertainty; ssL_C, sample-specific critical level sent to the National Water Information System (NWIS) database; *a priori* MDC, contractual *a priori* MDC; ssMDC, sample-specific Minimum Detectable Concentration; Code, remark or value-qualifier code(s) sent to the NWIS database; R, nondetect, result less than sample-specific critical level;), ssMDC exceeded the *a priori* MDC; =, negative result may indicate potential negative bias; NWQL, National Water Quality Laboratory; D, detection; ±, plus or minus; <, less than; >, greater than; WSC, Water Science Center]

Unrounded concentration	Unrounded CSU	ssL_C	ssMDC	*a priori* MDC	Code	Decision processes	Rounded concentration and (CSU) sent to the NWIS database
−2.523	0.731	0.93	2.7	3.0	R =	Result < ssL_C; reasonable relation between CSU, ssL_C and ssMDC; ssMDC < *a priori* MDC. However, the result is unacceptable or requires careful qualification because the negative concentration > 1.65 × CSU. Remark and value-qualifier codes sent to NWIS. WSC scientists should search for patterns among any samples in this category.	−2.52 (0.73)
0.636	2.542	3.2	2.7	3.0		Result < ssL_{Cc} and ssMDC < *a priori* MDC. However, the result is unacceptable because of the unusually high CSU and because the ssL_C and ssMDC are too small relative to the CSU. NWQL may not accept this result for technical reasons and may request a reanalysis of this sample. NWQL will try to determine the cause of error by looking at other data and information.	Results are not sent to the NWIS database unless the problem is resolved
1.005	0.544	0.53	2.7	3.0		Result > ssL_C and ssMDC < a priori MDC. However, the result is unacceptable because the ssL_C is too small relative to the CSU. NWQL may not accept this result for technical reasons and may request a reanalysis of this sample. NWQL will try to determine the cause of error by looking at other data and information.	Results are not sent to the NWIS database unless the problem is resolved
10.783	4.204	8.9	19.7	1.0	>	Result > ssL_C and there is a reasonable relation between CSU, ssL_C, and ssMDC. However, the ssMDC > a priori MDC. Upon further review, NWQL determined that a small sample volume was used. Therefore, because the result is positive and reasonable for sample size and reanalysis is not possible because of the lack of sample, the result is recorded in the NWIS database. Value-qualifier code sent to NWIS.	10.8 (4.2)

3.3   Review of Negative Results

Analysis of a radiological sample produces a gross signal response that is related to the quantity of the radionuclide present. However, random measurement uncertainties will cause this signal to vary somewhat if the measurement is repeated. A nonzero signal may be produced even when no radionuclide is present. For this reason, the contract laboratory analyzes an instrument background or a blank sample (discrete from the blank used for quality-control purposes) and subtracts its signal from the gross signal to obtain the net signal. If the measurement process is under control (free from systematic bias) and a series of blanks were analyzed and the background signal subtracted from each measurement, the results should be evenly distributed above and below a zero concentration, with negative values in approximately one-half of the blanks (see fig. A1 in the Appendix). Therefore, negative results are possible due to the randomness of the measurement process. Nevertheless, this does not imply that there is negative radioactivity. Each calculated result will have an associated CSU, and thus a confidence interval can be calculated and interpreted. Sometimes the lower end of the confidence interval may be negative, meaning that the true concentration may not be different from zero.

In order to determine if a negative result is valid, it is compared to the lower 95 percent one-sided confidence interval. A negative result is considered valid if the magnitude of the negative result is $\leq$ 1.65 times the reported CSU (1.65 is the 95th percentile of the standard normal distribution). When the magnitude of the negative result is greater than 1.65 times the reported CSU, the result may be considered invalid because there is less than 5-percent probability that the result is from a blank or instrument background distribution (for example, with a zero mean value), indicating that the measurement process may not be in control (see examples in table 3). Typical reasons for invalid negative results include a nonrepresentative background or blank signal or an inaccurate determination of radionuclide interferences. An invalid negative result can be reported with the corresponding value-qualifier code (see section 3.4). A valid negative result can be reported as a nondetect concentration.

3.4   Assigning Remark and Value-Qualifier Codes

Remark and value-qualifier codes are assigned by NWQL and are included with results whenever additional information is needed for interpretation. The following remark and value-qualifier codes with their explanations can be used with radiological results. Only one remark code can be included with a radiological result, whereas up to three value-qualifier codes can be used. Remark and value-qualifier codes are assigned to the results during evaluation by the NWQL (see examples in table 3). Contractual acceptance criteria associated with the

remark and value-qualifier codes can be found at
http://wwwnwql.cr.usgs.gov/USGS/acu_contracts.html.

Remark code	Explanation
R	Nondetect, result below sample-specific critical level (ssL_C)

Value-qualifier	Explanation
(	Blank greater than the sample-specific critical level (ssL_C)
)	Sample-specific Minimum Detectable Concentration (ssMDC) is above the contractual *a priori* MDC
/	Matrix Spike (MS) recovery is outside of contractual acceptable range (see Glossary for definition of recovery)
@	Exceeded sample holding time
\	Laboratory Control Sample (LCS) recovery is outside of contractual acceptable range
~	Duplicates are not within the contractual acceptance limits
=	Negative result may indicate potential negative bias
^	Yield is outside of contractual acceptable range (see glossary for definition of yield)

3.5   Information Sent to the National Water Information System (NWIS) Database

The following information is sent to the NWIS database. The concentration, CSU, and ssL_C are reported in either pCi/L or pCi/g.

- Site-agency code
- Station-identification number
- Sample-collection date
- Sample-collection time
- Sample-collection end date
- Sample-collection end time
- Sample-medium code
- Parameter Code
- Rounded concentration
- Remark and value-qualifier code(s)
- Rounded Combined Standard Uncertainty (1-sigma)

• Sample-specific critical level

The NWIS database information also is transferred to NWISWeb to provide electronic access to radiological and other water-quality information through website *http://water-data.usgs.gov/nwis*.

3.6 Information in Detailed Data Packages

Additional information associated with the sample analysis is provided in a compact disk data package sent to the WSC to assist with the review of the laboratory and field QA sample results. Some of the information in the data package is not recorded in the NWIS database. The data package includes a narrative, data, sample information, and laboratory information sections as shown in the following list. The data section provides the result, CSU (1-sigma), ssMDC, *a priori* MDC, ssL_C, percent yield, aliquant size, and results for laboratory quality-control samples.

• Data report narrative (additional details specific to the analyses; for example, relative percent difference for duplicate samples)

• Data section
   – Sample summaries (client sample ID, location, matrix, laboratory sample ID, chain of custody, sample date and time, amount of sample received, and the WSC contact)
   – Sample batch QC summary (number of blanks, laboratory control samples, and duplicates)
   – Work summary (date collected, date received, date analyzed, date reviewed)
   – Method blank results
   – Laboratory control sample results
   – Matrix spike results
   – Duplicate results
   – Results by sample and method

• Analytical Services Request (ASR) form for each sample

• Radiological login sheet

4. Water Science Center Reviews of National Water Information System (NWIS) Database Results

Specific information from NWIS is needed to complete a thorough review of radiological results. For the radiological data corresponding to samples analyzed after March 1, 2003, the following list of alpha parameters should be retrieved from NWIS into a "by-result" table for review.

PCODE – Parameter code

PSNAM – Parameter abbreviated name

REMRK – Remark code; this will include any remark code stored with the result

VALUE – Result value; if retrieved with the no-rounding option, this will be the laboratory result

UNITS – Result unit of measure

QUAL1 – First value-qualifier code stored with the result

QUAL2 – Second value-qualifier code stored with the result

QUAL3 – Third value-qualifier code stored with the result

LSDEV – Laboratory standard deviation; this field is where the Combined Standard Uncertainty (1σ CSU) for the result is stored

RLTYP – Report level type; for radiological samples analyzed after March 1, 2003, this field always equals "ssL_C"

RPLEV – Report level; this field is where the sample-specific critical level (ssL_C) for the result is stored

RCMLB – Result-level laboratory comment; this field will provide any additional information stored with the result

Results should be retrieved using the unrounded option because radiological results stored in NWIS are already rounded. The VALUE, LSDEV, and RPLEV are reported in the same UNITS.

For radiological samples analyzed before March 1, 2003, sample-specific critical levels (ssL_C) were not reported, and 2-sigma precision estimates or 2SPE (equivalent to 2 sigma Combined Standard Uncertainty) were reported under separate parameter codes. For tritium and radon samples analyzed prior to August 1, 2008, the ssL_C was not reported and the 2SPE was reported under a separate parameter code. For tritium and radon samples submitted after August 1, 2008, the ssL_C and 1σ CSU will be reported. More details about the retrieving results from the NWIS database can be found at *http://wwwnwql.cr.usgs.gov/USGS/rapi-note/05-019.html* and in section 3.4.6 of Web page *http://wwwnwis.er.usgs.gov/currentdocs/qw/QW.user.book.html*.

Much of the review by the WSC is focused on data interpretation. The WSC shall review the radiological results with respect to historical data from the collection site. Results obtained for QC samples, such as matrix-spike samples, can be reviewed to identify quality problems in laboratory analytical performance, sample matrix effects, and field sample collection. Matrix spike results can be used to establish bias, whereas laboratory-duplicate results can be used to establish subsampling and method variability. Duplicate field samples can be used to establish sample-collection variability. Remark and value-qualifier codes should be reviewed in order to evaluate their effect on interpretation and for providing descriptive information presented in publications.

5. Publishing Results

5.1 Technical Reports

The USGS conventions for publishing radiological results as outlined in this report follow the practices of the U.S. Environmental Protection Agency (U.S. Environmental Protection Agency, 1980b), American National Standards Institute N42.23 (American National Standards Institute, 2003), and Multi-Agency Radiological Laboratory Analytical Protocols (MARLAP, 2004, chapter 16). These national standards and guidance documents state that reported radiological data should always consist of two numbers, the measured concentration (or activity) and the associated measurement uncertainty (CSU at a stated level of confidence). Therefore, after radiological results have been reviewed by the WSC, the minimal acceptable information to be published shall include the:

- Result (positive, negative, or zero)

- CSU (1σ)

- Reference radionuclide for gross alpha and beta analyses

The concentration (or activity) and CSU should not be interpreted as a single point, but as a confidence interval about the measured concentration in which one has a high statistical probability of finding the true concentration of the sample (approximately 68 percent at 1-sigma). The practice of not including the CSU is ill advised as it withholds critical information associated with the result that could lead to misinterpretation or even critical misapplication of the data. Although the measurement uncertainty is not used in determining compliance with the Safe Drinking Water Act (SDWA), it will be needed for data evaluation of other studies.

The concentration, including zero and negative results, and the CSU shall be recorded in the same units (for example, picocuries per liter). In addition, the CSU shall never be stated as a relative percentage or fraction of the result because as a result approaches zero the relative uncertainty becomes exceedingly large and does not lend itself to meaningful interpretation. When nondetect results are published, it is strongly recommended that the report include a "detection indicator" for clarification. For example, the result and CSU are reported and flagged with a nondetect identifier whose definition is provided. Results shall not be reported as $<$ssMDC or $<$ssL$_C$.

Radiological results should be reported according to conventions that establish and preserve their technical defensibility. Results should be identified in a manner that permits them to be connected unambiguously to a sampling event

and to the radioanalytical measurements used to generate the results. Individual results are best reported in association with a unique identifier that is, or can be, associated with a project, location, date, time, and record of collection. If groups of data are being averaged, it may not be feasible to reference each unique identifier, but the descriptor associated with the averaged data point should always accurately and unambiguously characterize the group of data in question. Results should always be presented in association with the name of the analyte, the measured concentration (inclusive of all positive, zero, or negative values) and associated CSU, the level of significance for the confidence interval reported, the ssL$_C$, and an activity reference date for shorter lived radionuclides or mixtures of radionuclides. In the case of nonradionuclide-specific measurements, one should include the measurement parameter, such as gross alpha or total uranium, as well as any applicable assumptions underlying the gross measurement. For example, for gross alpha, the WSC should specify the reference nuclide used for calibration of the instrument that is "gross alpha (referenced to ^{230}Th)."

Oftentimes, the activity reference date and time are overlooked by investigators who are unfamiliar with radioanalytical measurements. The measured activity reported for a sample is only valid for a specified point in time because the radioactivity of a sample changes over time, depending on the half-life of the supporting radionuclide. Failure to specify the activity reference date and time, especially with short-lived radionuclides, can render published results useless. If the holding time for a sample analyzed for a short-lived radionuclide is exceeded, the published result shall include a statement that the holding time was exceeded with the specific time interval beyond the holding-time limit.

Table 4 shows an example of the types of information that should be included when publishing radiological results, such as those discussed in table 3. NWIS remark and value-qualifier codes can be translated to convey additional interpretive information for the data presented.

5.2 Nontechnical Reports

Presenting radiological data in a technically defensible, yet understandable manner is a challenge to any investigator who prepares reports for issuance to the general public. Clearly, WSCs must always ensure that data are presented in a manner that addresses the subject clearly without compromising technical accuracy or validity. While attempting to prevent or minimize confusion among the lay reader, the WSC may decrease the level of detail of the data provided or simplify the complexity of concepts presented to a level appropriate for the purpose and the perceived background of the audience. The WSC shall ensure that reports are presented clearly and that the depth and limitations of the presentation are clear to any reader, ranging from the layperson to the expert. Although authors cannot foresee every use or interpretation of their published data, it is important that they remain mindful that

Table 4. An example of typical information that should be provided when publishing radiological results.

[Result, radiological concentration plus or minus the 1-sigma combined standard uncertainty; ssL$_c$, sample-specific critical level; GA, gross alpha; 72h, sample analyzed for GA concentration at approximately 72 hours after sample collection as referenced to a detector calibrated using ^{230}Th; 30d, sample used for the 72-hour GA analysis is counted a second time approximately 30 days after the initial count as referenced to a detector calibrated using ^{230}Th; pCi/L, picocurie per liter; D, analyte detected; ND, analyte not detected, concentration is less than the sample-specific critical level; a, 72-hour sample holding time was exceeded; b, negative result may indicate potential negative bias; c, sample-specific Method Detectable Concentration (ssMDC) exceeded the *a priori* MDC; d, ssMDC exceeded the *a priori* MDC but was reasonable for the sample volume processed; ±, plus or minus; EPA, U.S. Environmental Protection Agency]

Sampling site	Sample collection date and time	Analysis date and time	Radiological constituent	Result	ssL$_c$	Remark	Units	Sample type	Analytical method
Well 5	9/18/2006 14:10	9/25/2006 8:35	GA (72h)	17 ± 12	23	ND, a	pCi/L	Unfiltered	EPA 900.0
Well 6	9/21/2006 8 37	9/23/2006 9:40	GA (72h)	5.1 ± 1.9	2.5	D	pCi/L	Filtered	EPA 900.0
Well 5	9/18/2006 14:10	10/20/2006 15:11	GA (30d)	23 ± 14	28	ND	pCi/L	Unfiltered	EPA 900.0
Well 6	9/21/2006 8 37	10/27/2006 8:25	GA (30d)	5.8 ± 2.7	4.2	D	pCi/L	Filtered	EPA 900.0
Well 5	10/7/2006 9 11	10/9/2006 8:11	^{228}Ra	2.35 ± 0.54	0.93	D	pCi/L	Filtered	EPA 904.0
Well 5	10/14/2006 9:15	11/16/2006 13:25	^{228}Ra	0.53 ± 0.54	0.93	ND	pCi/L	Filtered	EPA 904.0
Well 5	11/20/2006 9:04	11/25/2006 10:04	^{228}Ra	-2.52 ± 0.73	0.93	ND, b	pCi/L	Filtered	EPA 904.0
Well 5	10/22/2006 9:45	10/25/2006 14:45	^{228}Ra	6.6 ± 1.5	3.2	D, c	pCi/L	Filtered	EPA 904.0
Well 6	10/31/2006 9:30	11/2/2006 10:30	^{228}Ra	0.53 ± 0.74	1.6	ND, c	pCi/L	Filtered	EPA 904.0
Well 6	11/20/2006 9:10	11/25/2007 11:04	^{228}Ra	7.8 ± 2.2	3.9	D, d	pCi/L	Filtered	EPA 904.0

data may lose validity when it is taken out of the context or presented in an otherwise incomplete manner. WSCs shall always attempt to minimize the probability that results could be misinterpreted or misconstrued.

USGS WSCs conduct research studies and monitoring programs that focus on the detection and quantification of radiological constituents in various environmental matrices at substantially lower concentrations than those associated with regulatory action levels (AL), Safe Drinking Water Act Maximum Contaminant Levels (MCL), or other health benchmark levels. Surface water and ground water may reasonably be expected to have at least small amounts of some radionuclides. However, their presence does not necessarily indicate the water poses a health risk. Therefore, providing a comparison of WSC results to AL or MCL should be considered to ensure the lay public interprets the radiological concentrations from a relevant perspective. In addition, it also should be emphasized that nondetection does not imply that the radiological constituent is not present; rather, its concentration is below the level that can be measured. Table 5 shows an example of the types of information that should be included when publishing radiological results in nontechnical reports.

6. Interpretation and Reporting of Results from an Aggregated Dataset

As discussed in section 5, reporting of radiological results can be either simple and straightforward or challenging. Therefore, the aggregation of individual results into a single dataset for graphical presentation, summarization, or other purposes must be considered carefully within the limitations of individual results. For example, it is not uncommon to have an aggregated dataset that includes positive, negative, and zero results. The WSC should exercise caution when summarizing large-scale multisite, single-measurement datasets that have a large percentage of data below detection because such data could impart substantial weight to the overall statistical computation, depending on the treatment used. Appropriate statistical tools for the analysis of such datasets are presented by Helsel (2005) and Taylor (1990). Many of the same treatments that are used with other aggregated water-quality datasets are appropriate for aggregated radiological data as long as the implications and limitations cited in this document are clearly accounted for.

Table 5. An example of typical information that could be provided when publishing radiological results in a nontechnical report.

[Filtered water samples were collected from wells and analyzed for gross alpha, radium-226, and radium-228 using U.S. Environmental Protection Agency methods EPA 900.0, EPA 903.1 and EPA 904.0, respectively; gross alpha analysis is referenced to a detector calibrated using 230Th]

Sampling site	Contaminant, units[1]	MCL[1]	Number of samples[1]	Average concentration[1]	Number of results greater than the critical level[1]	Range of concentrations[1]
Well 5	Gross alpha, pCi/L	15	5	5.86	4	ND to 9.71
Well 6	Gross alpha, pCi/L	15	7	9.2	7	6.5 to 11
Well 5	^{226}Ra + ^{228}Ra, pCi/L	5	5	2.97	5	2.63 to 3.31
Well 6	^{226}Ra + ^{228}Ra, pCi/L	5	7	0.65	3	ND to 1.05

[1]Definitions:

The average contaminant concentration, uncertainty, and critical level for a radiological constituent(s) is given in picocuries per liter (pCi/L).

The average is calculated by adding together all the individual results from a sampling site and dividing the sum by the number of individual results.

The Safe Drinking Water Act's Maximum Contaminant Level (MCL) is the highest concentration of a contaminant that is allowed in drinking water.

The number of samples corresponds to the number of samples analyzed from the location.

The uncertainty characterizes the range of the concentrations, low to high limit, which could reasonably be attributed to the radiological measurement.

The critical level is the concentration below which results are considered to be nondetections with a 5-percent probability of false detection.

A radiological contaminant is not detected (ND) when its concentration is less than the critical level.

The combined standard uncertainty (CSU) may be used when interpreting radiological results. For example, a graphic display of the sample result and CSU for four samples collected from the same location is provided in figure 1. The

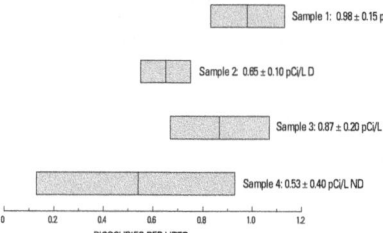

Figure 1.   Graphical interpretation of radiological results. The detection (D) and nondetection (ND) values are shown, and the 68-percent confidence level or 1-sigma Combined Standard Uncertainty (1σ CSU) are identified by the shaded areas. Units are picocuries per liter (pCi/L).

CSU provides an upper and lower limit to the range in which the true sample result lies; the bar chart shows the relation between the activities measured in individual samples.

The results and associated CSUs from two samples collected at the same site and time (or duplicate samples in a laboratory's batch QC) can be evaluated to determine whether they are statistically the same or different. An example of a simple equation that may be used to determine if two results (R_1 and R_2) with their associated CSUs (CSU_1 and CSU_2) are different is provided in equation 5. This equation is taken from the concept of normalized absolute difference (Paar and Porterfield, 1997), which tests the null hypothesis that the results do not differ significantly when compared to their respective CSUs. When the normalized absolute difference expression exceeds the z value, the results may be considered to be different on the basis of a defined significance level. It is common to use a z value of 2 or 3 (corresponding to 5 and about 0.3 percent significance levels, respectively).

$$| R_1 - R_2 | / \sqrt{\left(CSU_1^2 + CSU_2^2 \right)} > z \qquad (5)$$

Other possible approaches for interpreting or presenting aggregated radiological results, such as a statistical summary or graphical illustration, are provided herein as examples. These examples are not meant to be all-inclusive nor are they the only viable approaches. However, they serve to provide a perspective on aggregating and displaying radiological data. As with any interpretation or presentation of data, any approach should be reproducible and documented.

For certain projects, a WSC may want to use statistical analysis to summarize radiological results from samples collected from the same location at different times or from samples collected at different locations. Results obtained from the NWIS database should be reviewed and noted for acceptability before aggregating the data. When results are determined to be acceptable, the statistical analysis shall include the concentration and CSU (include the ssL_C for graphical representations) no matter if the result is negative, zero, or determined to be detected or nondetected. Excluding any positive, negative, or zero result from a dataset will bias the statistical evaluation and lead to possible erroneous conclusions.

Basic statistical terms such as the mean and standard error of the mean can be used to summarize aggregated measurements and are presented here. Other statistical approaches also can be used, but their description and use are beyond the scope of this report. Information on other statistical treatments can be found in Bevington and Robinson (1992). The average (x) of multiple laboratory measurements $x_1, x_2, ..., x_N$ of the same sample or of samples collected at the same location at the same time can be calculated using equation 6, where N is the number of measurements.

$$\overline{x} = \frac{x_1 + x_2 + \cdots + x_N}{N} \qquad (6)$$

The corresponding standard uncertainty of $\overline{x}$, based on the variance of the measurement $u^2(x_N)$, can be calculated using either equation 7 or 8, depending on how strongly correlated the measurements are with each other.

$$\frac{1}{N} \sqrt{u^2(x_1) + \cdots + u^2(x_N)} \qquad (7)$$

$$\frac{u(x_1) + \cdots + u(x_N)}{N} \qquad (8)$$

If all the measurement errors are essentially independent, the standard uncertainty is calculated using equation 7. If all the measurement errors are very strongly correlated, the standard uncertainty is calculated using equation 8. Equation 7 reduces the uncertainty roughly by a factor of $1/\sqrt{N}$, whereas equation 8 does not reduce the uncertainty at all.

For single measurements on samples collected at different locations and (or) times, it is not appropriate to propagate the uncertainties for the individual measurements when calculating the average because of the variability in sample collection. Sampling variability is usually assumed to be much larger than laboratory measurement variability. Therefore, in this case, the standard error of the mean is the best estimate of uncertainty for the average measurement and is calculated using equation 9.

$$s(\overline{x}) = \sqrt{\frac{1}{N(N-1)} \sum_{i=1}^{N} (x_i - \overline{x})^2} \qquad (9)$$

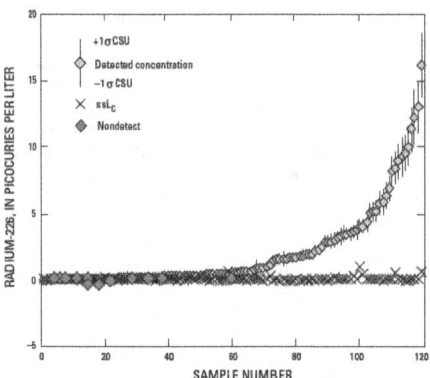

Figure 2. Graphical presentation of radiological results with pertinent supporting information.

All the required radiological parameters also can be presented graphically. Figure 2 illustrates ^{226}Ra results for 120 water samples. The figure includes reported results, their corresponding CSU and ssL$_C$, and nondetections. The dataset also included a few negative results that could be presented as reported and without manipulation or substitution. This simple illustration provides perspective on the overall dataset, including the range of detections and their concentrations. Using a graphical approach to presentation of data may facilitate interpretations not immediately evident in a data table by expanding the perspective on a large radiological dataset without censoring or manipulating the data.

Many USGS water-quality projects transcend the basic reporting of occurrence data described previously and require more detailed assessments of fate and transport or other process-oriented interpretations, such as comparisons of aggregated datasets between different environmental sites. It is beyond the scope of this report to provide details on all such activities. However, the WSC may build on the basic concepts of data reporting described in this report to reliably design and carry out its technically sound projects. Most of the potentially limiting factors are centered on the pitfalls that could arise if interpretations are made without consideration of the uncertainties in each individual result or aggregate of results. For example, in order to assess the fate of a naturally occurring radiological constituent, the WSC may need to distinguish between low levels of true environmental concentrations and the uncertainties due to laboratory analyses. In some cases, the uncertainties will be large, thereby making calculations of concentration gradients or differences between sites impossible or unreliable (for example, the data from sample numbers

1–40 in figure 2 are all of similar magnitude and associated uncertainty). On the other hand, in many cases, the differences between measurements will exceed the associated uncertainties by large margins and therefore could reliably be compared in these types of analyses (for example, in figure 2, the data point for sample number 80 could reliably be compared to the data point for sample number 100). In addition, statistical analysis of individual or aggregated data also should consider the associated uncertainties as part of the interpretation.

Acknowledgments

Anita Bhatt (U.S. Department of Energy-Idaho National Laboratory Radiological and Environmental Sciences Laboratory), Richard Graham (U.S. Environmental Protection Agency –Region VIII), John Griggs and Keith McCroan (U.S. Environmental Protection Agency –National Air and Radiation Environmental Laboratory), Rodney Melgard (Eberline Services Analytical Laboratory), Donovan Porterfield (Los Alamos National Laboratory), Robert Shannon (Quality Radioanalytical Support, LLC), and Myint Thein (U.S. Department of Energy, Oak Ridge National Laboratory) are thanked for graciously reviewing this report. The authors also thank James Eychaner, Michael Focazio, Dorrie Gellenbeck, Jeanne Hatcher, Gregory Mohrman, Lisa Olsen, Lisa Senior, Sylvia Stork, and Zoltan Szabo of the U.S. Geological Survey for their thorough reviews, which greatly improved the report.

References Cited

American National Standards Institute (ANSI), 2003, Measurement and associated instrumentation quality assurance for radioassay laboratories: American National Standards Institute Report N42.23.

Benjamin, J.R., and Cornell, C.A., 1970, Probability, statistics, and decision for civil engineers: New York, McGraw-Hill, 684 p.

Bevington, P.R., and Robinson, D.K., 1992, Data reduction and error analysis for the physical sciences: McGraw-Hill College, ISBN-13 978-0079112439.

Childress, C.J.O., Foreman, W.T., Conner, B.F., and Maloney, T.J., 1999, New reporting procedures based on long-term method detection levels and some considerations for interpretations of water-quality data provided by the U.S. Geological Survey National Water Quality Laboratory: U.S. Geological Survey Open-File Report 99–193, 19 p.

Currie, L.A., 1968, Limits for qualitative detection and quantitative determination—Application to radiochemistry: Analytical Chemistry, v. 40, p. 586–592.

Drever, J.I., 1988, The geochemistry of natural waters, 2d edition: Englewood Cliffs, New Jersey, Prentice-Hall, 437 p.

Eisenbud, M., and Gesell, T.F., 1997, Environmental radioactivity: Academic Press, 4th edition, ISBN-13, 978-0122351549.

Helsel, D.R., 2005, Nondetects and data analysis: Hoboken, New Jersey, John Wiley, 250 p.

Hem, J.D., 1985, Study and interpretation of the chemical characteristics of natural water, 3d edition: U.S. Geological Survey Water-Supply Paper 2254, 263 p.

International Organization for Standardization, 1995, Guide to the expression of uncertainty in measurement: International Organization for Standardization, Geneva, Switzerland.

International Union of Pure and Applied Chemistry, 1995, Nomenclature in evaluation of analytical methods including detection and quantification capabilities: Pure and Applied Chemistry, v. 67, no. 10, p. 1699–1723.

Lieser, K.H., 2001, Nuclear and radiochemistry—Fundamentals and application: Wiley-VCH, 2d edition, ISBN-13, 978-3527303175.

MARLAP (Multi-Agency Radiological Laboratory Analytical Protocols), NUREG-1576, 2004, U.S. Environmental Protection Agency publication number EPA 402-B-04-001C: National Technical Information Services publication number NTIS PB2004-106421, July edition.

Paar, J.G., and Porterfield, D.R., 1997, Evaluation of radiochemical data usability: U.S. Department of Energy, Office of Environmental Management, es/er/ms-5, 30 p.

Rucker, T.L., 2001, Calculation of decision levels and minimum detectable concentrations from method blank and sample uncertainty data—Utopian statistics: Journal of Radioanalytical and Nuclear Chemistry, v. 248, no. 1, p. 191–196.

Taylor, J.K., 1990, Statistical techniques for data analysis: Boca Raton, Fla., Lewis Publishers, Inc., ISBN 0-87371-250-1, 75 p.

U.S. Environmental Protection Agency, 1980a, Prescribed procedures for measurement of radioactivity in drinking water: U.S. Environmental Protection Agency, Radiochemical Methods Section, Physical and Chemical Methods Branch, Environmental Monitoring and Support Laboratory, Cincinnati, Ohio, publication number EPA 600/4-80-032.

U.S. Environmental Protection Agency, 1980b, Upgrading environmental radiation data: Health Physics Society Committee Report HPSR-1, U.S. Environmental Protection Agency publication number EPA 520/1-80-012, U.S. Environmental Protection Agency, Office of Radiation Programs, Washington, D.C.

U.S. Environmental Protection Agency, 2006, Inventory of radiological methodologies for sites contaminated with radioactive material: Office of Air and Radiation, National Air and Radiation Environmental Laboratory, Montgomery, Ala., U.S. Environmental Protection Agency publication number EPA 402-R-06-007.

U.S. Geological Survey, 2007, Policy for the evaluation and approval of analytical laboratories: Office of Water Quality Technical Memorandum No. 2007.01, accessed July 21, 2008, at *http://water.usgs.gov/admin/memo/QW/qw07.01. html.*

Glossary

Many of the definitions in the glossary are taken from MARLAP (2004).

A

activity: Mean rate of nuclear decay occurring in a given quantity of material. The "Curie" unit for activity is currently (2008) used by the USGS. The SI unit of activity is the becquerel (Bq), which equals one nuclear transformation per second. One curie equals 3.7×10^{10} Bq.

activity reference date: The date that is synonymous with the activity (concentration) on the day of sample collection.

aliquant: A representative portion of a homogeneous sample removed for the purpose of analysis or other chemical treatment. The quantity removed is not an evenly divisible part of the whole sample. By contrast, an aliquot is an evenly divisible part of the whole.

B

background (instrument): Radiation detected by an instrument when no source is present. The background radiation that is detected may come from radionuclides in the materials of construction of the detector, its housing, its electronics, and the building as well as the environment and natural radiation.

bias (of a measurement process): A persistent deviation of the mean measured result from the true or accepted reference value of the quantity being measured, which does not vary if a measurement is repeated.

blank (analytical or method): A sample that is assumed to be essentially free of the radionuclide that is carried through the radiochemical preparation, analysis, mounting, and measurement process in the same manner as a routine sample of a given matrix.

C

calibration: The set of operations that establish, under specified conditions, the relation between values indicated by a measuring instrument or measuring system, or values represented by a material measure, and the corresponding known value of a parameter of interest.

combined standard uncertainty (CSU): Standard uncertainty of an output estimate calculated by combining the standard uncertainties of the input estimates. The combined standard uncertainty of y is denoted by $u_c(y)$. The CSU is reported at the 68 percent or 1-sigma (1σ) confidence level.

critical level (L_C): In the context of analyte detection, critical level means the minimum measured value (for example, of the instrument signal or the radionuclide concentration) that indicates a positive (nonzero) amount of a radionuclide is present in the material within a specified probable error. The critical level is sometimes called the critical value or decision level. The general use of the term "critical level" is for a method wherein nominal measurement parameters are used in the calculation. Contrast this use to the sample-specific critical level (ssL_C) defined herein.

D

decay factor: The fractional amount of the original radionuclide activity in a sample that remains after decay in the time interval between sample collection and sample analysis.

duplicate sample: Two equal-sized samples of the material being analyzed, prepared, and analyzed separately as part of the same batch, used in the laboratory to measure the overall precision of the sample-measurement process beginning with laboratory subsampling of the field sample.

E

emission probability per decay event: The fraction of total decay events for which a particular particle or photon is emitted. The emission probability per decay event is also known as the branching fraction or branching ratio.

I

ingrowth factor: The activity of a supported radionuclide progeny at a specific time after chemical separation, expressed as a fraction of the amount of radioactivity at full ingrowth.

L

laboratory control sample: A standard material of known composition or an artificial sample (created by fortification of a clean material similar in nature to the sample), which is prepared and analyzed in the same manner as the sample. In an ideal situation, the result of an analysis of the laboratory control sample should be equivalent to (give 100 percent of) the target analyte concentration or activity known to be present in the fortified sample or standard material. The result normally is expressed as percent recovery.

M

matrix spike sample: An aliquant of a sample prepared by adding a known quantity of target analytes to a specified amount of matrix and subjected to the entire analytical procedure to establish if the method or procedure is appropriate for the analysis of the particular matrix.

maximum contaminant level (MCL): A regulatory limit established by the U.S. Environmental Protection Agency (USEPA) for the concentration of certain radionuclides in drinking water. The highest level (concentration) of a contaminant that is allowed in drinking water distributed to the public. MCLs are set as close as feasible to the level believed to cause no human health effects, while using the best available treatment technology and taking cost into consideration. MCLs are enforceable standards.

measurement quality objective (MQO): The analytical data requirements of the data-quality objectives, which are project- or program-specific and can be quantitative or qualitative. These analytical data requirements serve as performance measurement criteria or objectives of the analytical process. Multi-Agency Radiological Laboratory Analytical Protocols (MARLAP) refer to these performance objectives as MQOs. Examples of quantitative MQOs include statements of required radionuclide detectability and the uncertainty of the analytical protocol at a specified radionuclide concentration, such as an action level. Examples of qualitative MQOs include statements of the required specificity of the analytical protocol (for example, the ability to analyze for the radionuclide of interest, given the presence of interferences).

method blank: A sample assumed to be essentially target analyte-free that is carried through the radiochemical preparation, analysis, mounting, and measurement process in the same manner as a routine sample of a given matrix.

minimum detectable concentration (MDC): The minimum detectable concentration of the analyte in a sample. The smallest (true) radionuclide concentration that gives a specified probability (Type II - β) that the value of the measured radionuclide will exceed its critical level concentration (that is, that the material analyzed is not "blank" or free of analyte) (Currie, 1968; MARLAP, 2004, chapter 20). The general use of the term "MDC" or "*a priori* MDC" is for a method wherein nominal measurement parameters are used in the calculation. Contrast this use to the sample-specific MDC (ssMDC) defined herein.

N

nominal value: A value related to a designated or theoretical size that may vary from the actual.

null hypothesis (H_0): One of two mutually exclusive statements tested in a statistical hypothesis test (compare with alternative hypothesis). The null hypothesis is presumed to be true unless the test provides sufficient evidence to the contrary, in which case the null hypothesis is rejected and the alternative hypothesis (H_a) is accepted.

P

Poisson distribution: A random variable X has the Poisson distribution (Pr) with parameter λ if for any nonnegative integer k, $Pr[X = k] = (\lambda^k e^{-\lambda})/k!$. In this case, both the mean and variance of X are numerically equal to λ. The Poisson distribution is often used as a model for the result of a nuclear counting measurement.

precision: The closeness of agreement between independent test results obtained by applying the experimental procedure under stipulated conditions. Precision may be expressed as the standard deviation. Conversely, imprecision is the variation of the results in a set of replicate measurements.

Q

quality control (QC): The overall system of technical activities that measures the attributes and performance of a process, item, or service against defined standards to verify that they meet the stated requirements established by the project; operational techniques and activities that are used to fulfill requirements for quality. This system of activities and checks is used to ensure that measurement systems are maintained within prescribed limits, providing protection against out-of-control conditions and ensuring that the results are of acceptable quality.

quantile: A p-quantile of a random variable X is any value x_p such that the probability that X $< x_p$ is at most p and the probability that $X \leq x_p$ is at least p.

R

radioactivity: The property possessed by some elements or isotopes of spontaneously emitting energetic particles (electrons or alpha particles) by the disintegration of their atomic nuclei.

radioanalytical analysis: A general term used to denote the analysis of a sample for a specific radionuclide, group of radionuclides, or gross screening of radioactivity. The term may be used for a single radionuclide analysis or to denote a collection of analyses that may include gamma-ray spectrometric analyses, gross alpha and beta analyses, and specific radionuclide analyses that require chemical separations such as isotopic uranium, ^{226}Ra, and ^{90}Sr.

radiochemical analysis: A term used to denote the analysis of a radionuclide in a sample that requires chemical processes to isolate the radionuclide in the sample. Isotopic uranium, ^{226}Ra, and ^{90}Sr in a sample require radiochemical analyses.

radiological: An adjective relating to nuclear radiation.

radiological hold time: Refers to the time differential between the sample collection date and the final sample counting (analysis) date.

radionuclide: A nuclide that is radioactive (capable of undergoing radioactive decay).

recovery: The ratio of the amount of analyte measured in a spiked or laboratory control sample, to the amount of analyte added, usually expressed as a percentage. For a matrix spike, the measured amount of analyte is first

decreased by the measured amount of analyte in the sample that was present before spiking. Contrast this to yield defined herein.

relative standard uncertainty: The ratio of the standard uncertainty of a measured result to the result itself. The relative standard uncertainty of x may be denoted by $u_r(x)$.

S

sample: A portion of material selected from a larger quantity of material or a set of individual samples or measurements drawn from a population whose properties are studied to gain information about the entire population.

sample-specific critical level (ssL$_C$):   The sample-specific critical level is calculated using the parameter values measured during the generation of the sample result. This is different than the critical level for a method wherein nominal measurement parameters are used in the calculation (see critical level). Concentrations below the ssL$_C$ are considered nondetections.

sample-specific minimum detectable concentration (ssMDC):   The sample-specific minimum detectable concentration is calculated using the parameter values measured during the generation of the sample result (see minimum detectable concentration).

significance level: The risk (probability) of making a Type I error (α) is traditionally called the level of significance of the test.

standard uncertainty: The uncertainty of a measured value expressed as an estimated standard deviation, often call a "1-sigma" ($1-\sigma$) uncertainty. The standard uncertainty of a value x is denoted by $u(x)$.

T

Type I decision error: In a hypothesis test, the error made by rejecting the null hypothesis when it is true. A Type I decision error is sometimes called a "false detection" or a "false positive."

Type II decision error: In a hypothesis test, the error made by failing to reject the null hypothesis when it is false. A Type II decision error is sometimes called a "false nondetection" or a "false negative."

U

uncertainty: A parameter, usually associated with the result of a measurement, that characterizes the dispersion of the values that could reasonably be attributed to the measurement of interest.

Y

yield: The ratio of the amount of radiotracer or carrier determined in a sample analysis to the amount of radiotracer or carrier originally added to a sample. The yield is an estimate of the analyte during analytical processing. It is used as a correction factor to determine the amount of radionuclide (analyte) originally present in the sample. Yield is typically measured gravimetrically (through a carrier) or radiometrically (through a radiotracer).

Appendix: Typical Equations for Calculating Radiological Parameters

The information presented in this appendix conforms to the formulations and concepts presented in the Multi-Agency Radiological Laboratory Analytical Protocols, commonly referred to as MARLAP (2004), the International Organization for Standardization (1995), and International Union of Pure and Applied Chemistry (1995).

A1.   Calculating the Concentration

The general equation used to calculate concentration takes on the following form (MARLAP, 2004, chapter 19):

$$Concentration = \frac{(Gross\ Instrument\ Signal)\ -\ (Blank\ Signal + Estimated\ Interferences)}{(Sensitivity\ Factor)} \tag{A1}$$

where:

Concentration is in pCi/L;

Gross Instrument Signal is detector response in units of detector events registered per unit time from all constituents in the sample (includes detector and process-related background events);

Blank Signal is detector events registered per unit time from analyzing a blank sample having no target constituent;

Estimated Interferences is detector counts registered per unit time from nontarget constituents in the sample; and

Sensitivity Factor is a combination of multiplicative parameters, such as sample size, detector efficiency for radiation emitted, chemical yield of process, decay factor, ingrowth factor, and unit conversion factor.

The complexity of the equation used to calculate the concentration varies substantially and usually depends on the type of radiation emitted during radioactive decay (α, β, γ) and the radiochemical method chosen. The basic equation that has all possible parameters incorporated in a complex equation, such as for ^{228}Ra or ^{90}Sr by the analysis of their decay products ^{228}Ac and ^{90}Y, has the form taken from "Inventory of Radiological Methodologies for Sites Contaminated with Radioactive Material" (U.S. Environmental Protection Agency, 2006).

$$Concentration\ (pCi\ /\ L) = \frac{\left(\dfrac{N_s}{t_s} - \dfrac{N_B}{t_B}\right) \times \lambda_2 t_S}{CF \times B \times Y \times \varepsilon \times V \times I \times DF_P \times DF_{DP} \times \left(1 - e^{-\lambda_2 t_S}\right)} \tag{A2}$$

where:

N_S	is the number of accumulated detector events (counts) for the decay product;
N_B	is the number of background detector events (counts) for the equivalent sample count interval;
t	is the counting interval for sample (t_S) and background (t_B);
CF	is the factor used for converting to desired reporting units, typically 2.22 disintegrations per picocurie (pCi);
V	is the sample size (mass or volume);
B	is the branching fraction of the particle emission being counted. This is the fraction of all decays that result in an emission of the characteristic radiation (alpha, beta, or gamma);
Y	is the chemical yield of the analysis;
ε	is the detector efficiency for the particular emission of the radionuclide;
λ	is the decay constant of the radionuclide ($\lambda = 0.69315/t_{1/2}$; and $t_{1/2}$ is the half-life of the radionuclide (λ_1 for parent and λ_2 for the decay product);
DF_P or $\left(e^{-\lambda_1 T_1}\right)$	is the decay correction factor for the parent from sample collection to second chemical separation;
DF_{DP} or $\left(e^{-\lambda_2 T_3}\right)$	is the decay correction factor for decay product from second chemical separation to start of counting;
I or $\left(1 - e^{-\lambda_2 T_2}\right)$	is the ingrowth correction factor for the ingrowth of the decay product used to calculate the activity of a parent;
T_1	is the time interval between sampling and beginning the sample count;
T_2	is the time interval between first and second decay product chemical separations;
T_3	is the time interval between second decay product chemical separation and count;
$\lambda_2 t_s / (1 - e^{-\lambda_2 t_s})$	is the correction factor for radioactive decay of the decay product during the counting interval.

Note: Because measurement parameters are never truly known exactly, each factor has an associated standard uncertainty.

For a simple analysis, such as gross alpha and beta analysis, the following equation is used:

$$"c_\alpha" \text{ Concentration } (pCi/L) = \frac{\left(\dfrac{N_S}{t_S} - \dfrac{N_B}{t_B} \right)}{CF \times \varepsilon \times V} \tag{A3}$$

A2.   Calculating the Combined Standard Uncertainty

As presented in section A1, each factor in the equation used to calculate the concentration has an associated uncertainty called the standard uncertainty. By convention, a standard uncertainty is quoted at the 68-percent or 1-sigma confidence level. For example, the determination of the detector efficiency is not exact because of the uncertainty in the radioactive source used to determine the detector efficiency and the Poisson random counting uncertainty when counting the radioactive source.

The uncertainty for the stated concentration is calculated by combining (or propagating) these standard uncertainties into an overall uncertainty called the combined standard uncertainty. MARLAP (2004, chapter 19) uses a first-order uncertainty propagation formula shown in equation A4 to propagate the standard uncertainties. The general formula uses the partial differential of each factor (x_i) to calculate the concentration y. Equation A4 is used to calculate the variance $u_c^2(y)$ of the concentration based on the uncertainty propagation with uncorrelated inputs.

$$u_c^2(y) = \sum_{i=1}^{N} \left(\frac{\partial f}{\partial x_i} \right)^2 u^2(x_i) \tag{A4}$$

The combined standard uncertainty u_c (y) is calculated by taking the square root of the variance $u_c^2(y)$.

For the simple concentration equation A3, the resulting equation (MARLAP, 2004, chapter 19) used to calculate the variance is the following:

$$u_c^2(c_\alpha) = \left(\frac{\partial c_\alpha}{\partial N_S} \right)^2 u^2(N_S) + \left(\frac{\partial c_\alpha}{\partial N_B} \right)^2 u^2(N_B) + \left(\frac{\partial c_\alpha}{\partial \varepsilon} \right)^2 u^2(\varepsilon) + \left(\frac{\partial c_\alpha}{\partial V} \right)^2 u^2(V) \tag{A5}$$

where the uncertainties in the time variables are assumed to be negligible compared to the uncertainties of the other terms. Equation A5 reduces to equation A6 using the same parameters defined for equation A2.

$$u_c^2(c_\alpha) = \frac{\left(N_S \times t_B^2 - N_B \times t_S^2 \right)}{t_S^2 \times t_B^2 \times \varepsilon^2 \times V^2} + \frac{\left(\dfrac{N_S}{t_S} - \dfrac{N_B}{t_B} \right)}{\varepsilon^2 \times V^2} \times \left[\frac{u^2(\varepsilon)}{\varepsilon^2} + \frac{u^2(V)}{V^2} \right] \tag{A6}$$

The combined standard uncertainty $u_c(c_\alpha)$ of c_α is calculated by

$$u_c(c_\alpha) = \sqrt{u_c^2(c_\alpha)} \tag{A7}$$

A3.   Calculating the Critical Level

A3.1   General Principles

An analysis of a radiological sample will produce a gross signal response that is related to the quantity of analyte present in the sample. However, random measurement uncertainties will cause this signal to vary somewhat if the measurement is repeated on the same sample. A nonzero signal may be (and usually is) produced even when no analyte is present. For this reason, a laboratory analyzes a blank (or an instrument background) to determine the signal observed when no analyte is present in the blank sample and subtracts this blank signal from the gross signal to obtain a net signal for the sample being analyzed. In fact, because the signal varies when the blank measurement is repeated, there is a blank signal distribution whose parameters must be estimated. In a similar manner, the net signal response will have a distribution with an average value (assumed to be zero or μ_0) and a standard deviation (σ_0).

To determine how large an instrument signal must be to provide high confidence for the presence of the analyte in a sample, one calculates a threshold value for the net signal called the critical level L_C (Currie, 1968). The critical level also is denoted by S_C (MARLAP, 2004, chapter 20). If the observed net signal for a sample being analyzed exceeds the critical level, the radiological constituent is considered detected; otherwise, it is not detected. Because the measurement process is statistical in nature, it is possible for the net signal related to a blank sample to exceed the critical level, leading one to conclude incorrectly that the sample contains a positive amount of the analyte. Such an error is sometimes called a "false positive," although the term Type I error terminology is favored by MARLAP (2004, chapter 20). The probability of a Type I error is often denoted by α. Before calculating the critical level, one must choose a level for α. The most commonly used level is 0.05, or 5 percent. If $\alpha = 0.05$, then one expects the net instrument signal to exceed the critical level in only about 5 percent of cases (1 in 20) when analyte-free samples or blank samples are analyzed. These concepts are shown graphically in figure A1.

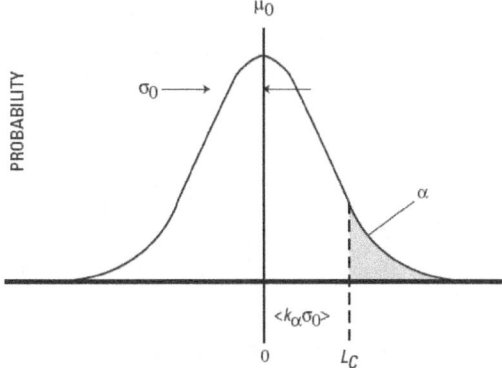

NET RESPONSE OF BLANK SAMPLES
(INSTRUMENT BACKGROUND SUBTRACTED)

Figure A1.   The curve representing the critical level concept by using a theoretical distribution of the net instrument signal (concentration) obtained when analyzing an analyte-free sample. The chosen Type I error probability σ determines the location of the critical level of the net signal L_C. For an $\alpha = 0.05$ or 5 percent, the critical level corresponds to the 95th quantile value of the normal distribution. The probability α is depicted as the area under the curve to the right of the dashed line. Note that decreasing the value of α requires increasing the critical value (shifting the dashed line to the right) and increasing the value of α decreasing the critical level (shifting the dashed line to the left). The α quantile of the standard normal distribution, using a default level of 1.645, is k_α; the mean of the net signal responses of blank sample distribution is μ_0; and the standard deviation of the net signal response of blank sample distribution is σ_0.

A3.2   Calculating the Sample-Specific Critical Level (ssL$_c$)

The sample-specific critical level (ssL$_C$) is calculated by the contract laboratory using all the sample-specific parameter values. A sample-specific critical level is calculated using the net instrument background counting distribution and sample-specific factors, and by assuming that the mean value of the net instrument background distribution is zero. The ssL$_C$ is commonly calculated using equations A8a and A8b (Currie, 1968).

$$ssL_C = \frac{1.645 \times \sigma_0}{CF \times t_s \times V \times B \times \varepsilon \times Y \times DF \times I} \tag{A8a}$$

and

$$\sigma_0 = \sqrt{N_B \frac{t_S}{t_B} \left(1 + \frac{t_S}{t_B}\right)} \tag{A8b}$$

where:

σ_0 is the standard deviation of the net instrument background (counts);
N_B is the background counts in background counting interval
CF is the unit conversion factor, typically 2.22 disintegrations per minute per picocurie;
t_S is the counting time of sample (minute);
t_B is the counting time of background (minute);
V is the sample size (mass or volume);
B is the branching fraction of the particle emission being counted;
ε is the fractional detector efficiency for the particular emission of the radionuclide;
Y is the chemical yield of the analysis;
DF is the decay factor; and
I is the correction factor for the ingrowth of progeny used to calculate the activity of a parent.

Equation A8a is not suitable when the number of detector background counts is small or zero over the background counting time. This may be the case for analyses that use low background detection systems, such as alpha or gamma ray spectrometers. For such applications, the equations discussed in chapter 20, attachment 20A of MARLAP (2004) need to be reviewed and applied.

The specific equation that is used by the contract laboratories to calculate the ssL_c for various radioanalytical methods will be specified by the NWQL in the PWS.

A3.3   Practical Approach for Verifying the Reported Sample-Specific Critical Level

From a practical point of view, when the sample and background counting times are nearly equal, multiplying the CSU associated with a QC blank's result by 2 approximates the critical level of a method (not an individual result). However, this approximation is only good when the parameters used to calculate the concentration for the blank are nearly equal to the average values used for the method.

An approximation of 2 times the CSU of the sample or blank result (or sometimes slightly greater depending on the background of the instrument used for the measurement) can be used for verifying the reported ssL_C. This assumes near equal sample and background counting times. For example, if a radiological result is 0.121 ± 0.091 pCi/L, the ssL_C can be estimated to be 0.18 or about 0.20 pCi/L. This ssL_C approximation is useful to verify that the laboratory has not made a substantial error in calculating the sample-specific critical level. The approximation of 2 times the CSU may provide reasonable estimates for most applications but should not be used to categorically indicate that a mistake has been made without evaluating the relation between the reported result, CSU, ssL_C, and ssMDC.

A4.   Calculating the Minimum Detectable Concentration

A4.1   General Principles

The *a priori* minimum detectable concentration (*a priori* MDC) is a hypothetical predictive concept that is used to estimate the detection capability of a measurement process (method) under defined circumstances. The concept is *a priori* or before the fact (before a sample measurement) and is not to be used to evaluate individual measurement results. After a measurement has been made, the result and its Combined Standard Uncertainty (CSU) are the important quantities that are used to compare analytical results to historical values for the sampling site.

Once the critical level L_C (instrument response) has been defined for a method based on a distribution of blank samples and an α probability of Type I error (false detection) rate, an *a priori* MDC may be established by specifying the acceptable Type II error rate β (false nondetection) and the standard deviation of the probability distribution σ_D of the net signal response when the true value S_D is equal to the MDC value (Currie, 1968). The concept assumes no systematic errors such as method bias. The *a priori* MDC is defined so that the probability distribution of the possible measurement responses when S_D equals the MDC crosses the critical level L_C at the 1-β fraction of the distribution.

Graphically, the theoretical *a priori* MDC distribution and its relation to the critical level are shown in figure A2 (based on figure 20.1 in MARLAP, 2004). In this figure, S_D corresponds to the *a priori* MDC value. Under the *a priori* MDC concept, the shaded area under the S_D distribution corresponds to the β or the lower 5 percent of the distribution. Analytical results with values in the shaded area would not be considered different from background because their values are below the critical level. The nonshaded area of the S_D distribution corresponds to the 1-β or the upper 95 percent of the distribution. Analytical results with values in this region would be considered positive because their values are greater than the critical level. Note that all possible analytical result values between L_C and S_D (45 percent of the S_D distribution) would be considered different from background and thus positive values.

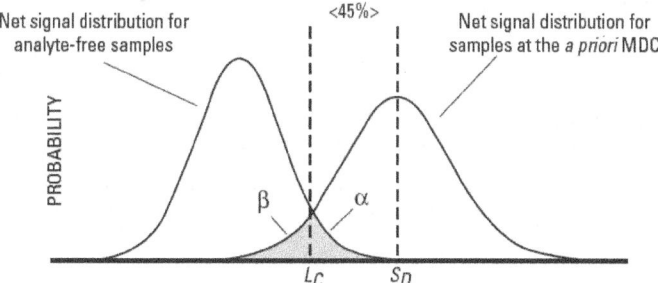

Figure A2. Graphical representation of the *a priori* Minimum Detectable Concentration concept (taken from figure 20.1; in MARLAP, 2004). The false nondetection Type I error (α, 5-percent probability), the false detection Type II error (β, 5-percent probability), the critical level (L_c), and the *a priori* minimum detectable concentration (S_D) are shown in the figure.

If each sample in a set of 100 samples were spiked with a radionuclide at the *a priori* MDC concentration and analyzed, the mean concentration of the 100 analyses would be the *a priori* MDC value (S_D) and the relative one standard deviation of the distribution of analytical results around the *a priori* MDC would be about 30 percent. Therefore, the spread in the individual results from the set of 100 samples would range from slightly less (about 5 percent) than the critical level to about twice the *a priori* MDC value. However, 95 out of the 100 results would exceed the critical level. By specifying the *a priori* MDC in the PWS, a false nondetection β probability of 5 percent could be assumed for a distribution of results (from multiple samples or analyses of the same sample) when the true concentration is at the hypothetical MDC. For USGS applications, typically only one sample from a particular site may be provided and only one analysis of the sample performed. A result from a single analysis that is less than the hypothetical *a priori* MDC may be from the MDC distribution, but it also could be from distributions whose true concentration is less than or greater than the MDC. When a single analysis of an individual sample is made, the true concentration in the sample can be estimated with approximately 95-percent confidence to be within the range described by the reported value ± 2 CSU.

Mathematically, the *a priori* MDC S_D can be calculated using many equations, each having a different assumption and approach. MARLAP (2004, chapter 20) discusses these different approaches. The most simplified approach assumes a Gaussian distribution for L_C and S_D instrument responses. Currie (1968) used a version of the following simplified equation for instrument response, not concentration:

$$S_D = L_C + k_\beta \times \sigma_D \qquad (A9)$$

where:

k_β is the $1 - \beta$ quantile of the standard normal distribution, default value of 1.645;
L_C is the critical level response, defined earlier; and

σ_D is the standard deviation of the S_D distribution (net response).

When $\alpha=\beta$, $k_\alpha=k_\beta$, and $\sigma_D \sim \sigma_0$ and where σ_0 is the standard deviation of the net background response distribution, equation A9 reduces to the most simplified form:

$$S_D = 2.71 + 3.29 \times \sigma_0 \qquad (A10)$$

When the net response background distribution is more like a Poisson distribution (when the background or sample counts for the analysis are less than 70) rather than like the assumed Gaussian distribution, the use of this equation may give an MDC wherein the observed probability of false detection may be higher than the assumed α value.

An approach has been suggested by Rucker (2001) for determining a method *a priori* MDC-based net background standard deviation S_{B0}, in picocuries per liter, from a population of blank sample results (instrument background subtracted) as a substitute for use in equations A8 and A11. In addition, the k_α and k_β for equations A10 and A11 are replaced with the Student's t factor for the appropriate number of degrees of freedom. The MDC equation proposed by Rucker is

$$MDC = t^2 + 2 \times t \times S_{B0} \qquad (A11)$$

where:
S_{B0} is the standard deviation of the distribution net blank results in picocuries per liter; and
t is the Student's t factor for the number of blank samples used to determine S_B and for the default α and β probabilities of 0.05.

As noted by Rucker, this approach accounts for all of the uncertainty in the measurements (due to the variability in the parameter values used to calculate the result), not just the counting uncertainty. As such, this approach is useful in estimating the *a priori* MDC for a method, not a sample-specific MDC, and can be applied to defining the sensitivity requirements for contract laboratory work.

A4.2   Calculating the Sample-Specific Minimum Detectable Concentration (ssMDC)

Some laboratories calculate the sample-specific Minimum Detectable Concentration (ssMDC) based on the instrument background and applicable sample-specific parameters according to the following general equation:

$$ssMDC = \frac{3.29 \times \sigma_0 + 2.71}{CF \times t_s \times B \times Y \times \varepsilon \times V \times DF \times I} \qquad (A12a)$$

and

$$\sigma_0 = \sqrt{N_B \frac{t_s}{t_B}\left(1 + \frac{t_s}{t_B}\right)} \qquad (A12b)$$

where:
σ_0 is the standard deviation of the net instrument background (counts);
N_B is the background counts in background counting interval;
CF is the unit conversion factor, typically 2.22 disintegrations per picocurie;
t_S is the counting time of sample (minute);
t_B is the counting time of background (minute);
V is the sample size (mass or volume);
B is the branching fraction of the particle emission being counted;
ε is the fractional detector efficiency for the particular emission of the radionuclide;
Y is the chemical yield of the analysis;
DF is the decay factor; and
I is the correction factor for the ingrowth of progeny used to calculate the activity of a parent.

Equations A12a and A12b are not used when the number of detector background counts is small or zero over the background counting time. This may be the case for analyses that use low background detection systems, such as alpha- or gamma-ray spectrometers. For such applications, the equations discussed by MARLAP (2004, chapter 20, attachment 20A) need to be reviewed and applied.

In the original concept, the average values of the applicable parameters were used to calculate the *a priori* MDC for a method. However, more recently, sample-specific values of the parameters are used to calculate the ssMDC to show that the laboratory has met the required *a priori* MDC specified in a contract. The ssMDC is not used to determine if results are different from an instrument background or a blank sample.

The specific equation that is used by the contract laboratories to calculate the ssMDC for various radioanalytical methods will be specified in the PWS by the NWQL.

A4.3 Practical Approach for Verifying the Reported Sample-Specific Minimum Detectable Concentration

There are at least two approaches or rules of thumb that can be used to estimate whether a reported ssMDC has been calculated properly. The first approach compares the reported ssMDC to the reported ssL_C. The ssMDC should be approximately 2 times the reported ssL_C (or slightly greater depending on the background of the instrument used for the measurement). The second approach is to compare the reported ssMDC of a sample to a multiple of the reported CSU. This approach is only applicable when dealing with acceptable blank and negative results or with acceptable positive results whose absolute value is less than approximately 3 times the reported ssMDC value. For such cases, the reported ssMDC should be approximately 3 to 4 times the CSU of the sample result. Different multipliers may be applied for certain methods of analysis and instrument backgrounds. Analytical concentrations at or near the ssMDC should have a relative CSU of approximately 30 percent for most methods, with the exception of methods for the low-level determination of alpha-emitting nuclides. The relative CSU decreases with successively higher concentrations above the MDC. These rules of thumb are based on the assumed relation of the relative CSU at the MDC for paired observations.

A4.4 The Effect of Sample Size and Counting Time on the Reported Sample-Specific Minimum Detectable Concentration (ssMDC)

The magnitude of the ssMDC is inversely proportional to the sample volume analyzed and inversely proportional to the square root of the counting time (see equations A12a and A12b in section A4.2). Figure A3 shows the effect of reducing the size of a 1-liter sample on the magnitude of the ssMDC. When the sample volume analyzed is reduced by a factor of 2 (1 L to 0.5 L), the magnitude of ssMDC is increased by a factor of 2 (1 pCi/L to 2 pCi/L). Figure A4 shows the effect of reducing the amount of time a sample is analyzed by a radiation detector (instrument counting time) on the magnitude of the ssMDC. When the counting time is reduced by a factor of 2 from 200 minutes to 100 minutes, the magnitude of the MDC is increased by a factor of $\sqrt{2}$ from 1 pCi/L to 1.41 pCi/L (see equations A12a and A12b).

A4.5 The Relation between the Combined Standard Uncertainty and the Calculated Activity in a Sample for Two Radioanalytical Measurement Techniques

Several graphical illustrations of the relation between the reported activity in a sample and the Combined Standard Uncertainty (CSU) of the result have been developed. These relations are shown in figures A5a and A5b for gross beta activity analysis of water and in figures A6a and A6b for isotopic uranium analysis of water by alpha spectrometry techniques. Typical sources (detector efficiency and chemical yield) and their standard uncertainties were used (for illustrative purposes only) in the calculation of the CSU and the development of the graphs.

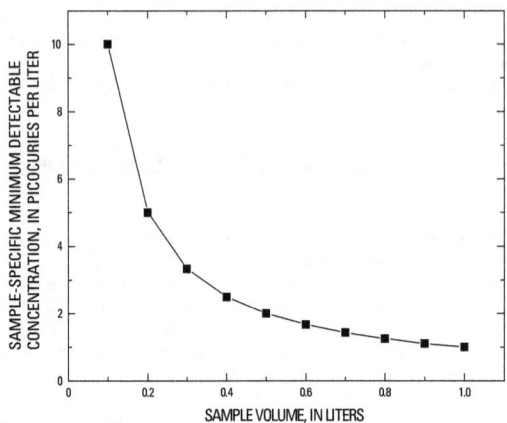

Figure A3.   The sample-specific Minimum Detectable Concentration
(ssMDC) as a function of analysis sample volume.

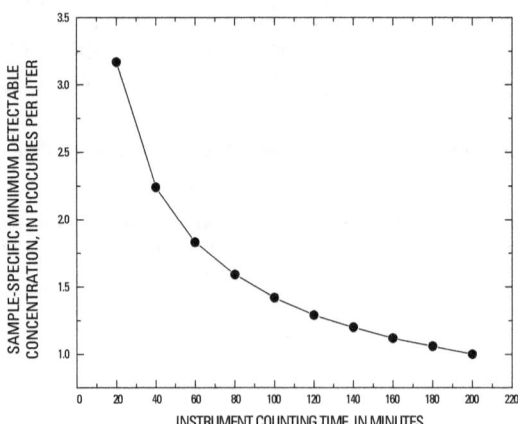

Figure A4.   The sample-specific Minimum Detectable Concentration
(ssMDC) as a function of analysis counting time.

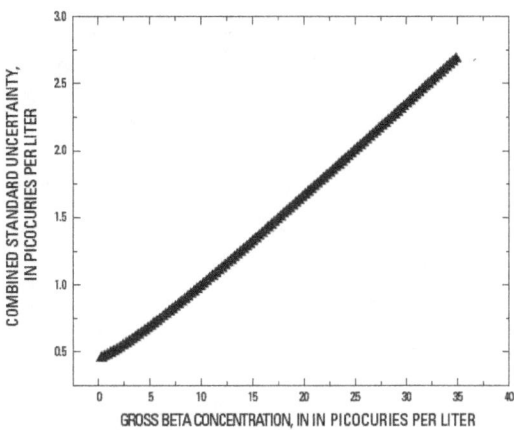

Figure A5a.   Typical Combined Standard Uncertainty (CSU) as a function of gross beta concentration.

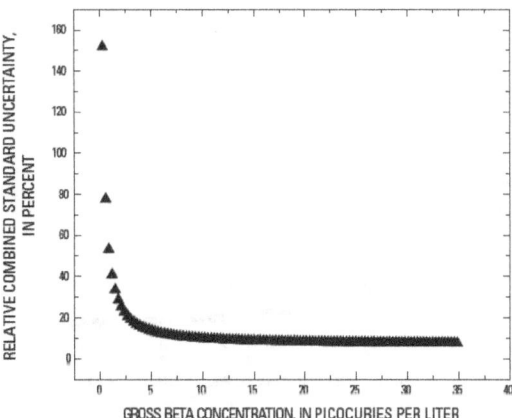

Figure A5b.   Typical relative Combined Standard Uncertainty (CSU) as a function of gross beta concentration.

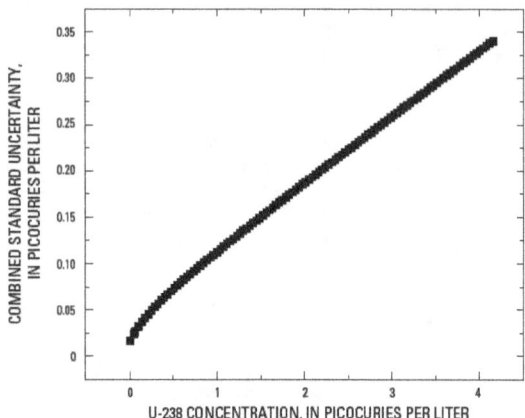

Figure A6a. Typical Combined Standard Uncertainty (CSU) as a function of Uranium-238 concentration.

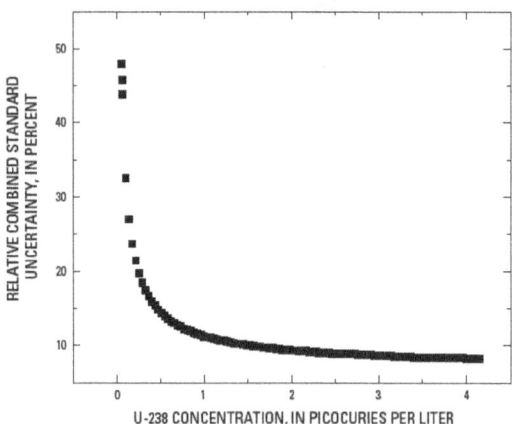

Figure A6b. Typical relative Combined Standard Uncertainty (CSU) as a function of Uranium-238 concentration.

www.ingramcontent.com/pod-product-compliance
Lightning Source LLC
Chambersburg PA
CBHW071551170526
45166CB00004B/1635